RISE and STAND FIRM

Linda Williamson

Defender Publications
Harrison, TN

Published by Defender Publications, Harrison, TN, USA

Copyright 2020 Linda Williamson

All rights reserved. No part of this book may be used or reproduced in any manner whatsoever without written permission, except in the case of brief quotations in critical articles and reviews. For more information, contact Defender Publications:
SpecialSales@RBEnterprises.info

First edition printed 2020
ISBN: 978-1-950038-13-8 (paperback)
ISBN: 978-1-950038-14-5 (eBook: ePUB)

PUBLISHERS NOTE:

The scanning, uploading, and distribution of this book via the Internet or via any other means without the permission of the publisher is illegal and punishable by law. Please purchase only authorized print and/or electronic editions, and do not participate in or encourage piracy of copyrighted materials. Your support of the author's rights is appreciated.
The publisher does not have any control over and does not assume any responsibility for author or third-party web sites or their content.
Defender Publications books are available at special discounts for bulk purchases, for sales promotions, or corporate use. Special editions, including personalized covers, excerpts of existing books, or books with corporate logos, can be created for some titles. For more information, contact Defender Publications at:
SpecialSales@RBEnterprises.info

Table of Contents

Preface 5

Beginnings 7

The Folder 11

Tears in a Bottle 13

The Routine 33

The Crucible 41

The Reunion 47

Preface

"Be strong; be brave; and always love Jesus." As a young mom, those were the words I voiced repeatedly to my growing son, never realizing how literally our children accept the things we speak over them. The day came too soon when he decided to actually be strong and brave and join the United States Marines; and this mom was not at all prepared. If you are or will be the parent of a Marine, your events may be different than mine. My voyage was laced with fear and anxiety. Yours may be joyful and excited, or somewhere in between. But our journey will take us to the same outcome: pride. Regardless of the road we take to get there, we all wind up staring at those yellow footprints, amazed that something we once cradled tightly in our protective arms is now standing before us as a United States Marine, ready and willing to be strong and brave for us all.

When my kids were little, I prayed Psalms 19 and 20 over them while they were at school. Now I pray these words over their futures:

> *"May the words of their mouth and the meditation of their heart be acceptable to you, Lord, my Rock and my Redeemer. May the Lord answer you in a day of trouble; may the name of Jacob's God protect you. May he send you help from the sanctuary and sustain you. May he give you what your heart desires and fulfill your whole purpose. Now I know that the Lord gives victory to his anointed; he will answer from his holy heaven. We take pride in the name of the Lord our God. We rise and stand firm".*

This is the story of how my son chose to Rise and Stand Firm.

Linda Williamson, Author

Beginnings

My story begins on March 23, 2015, but my journey to this point started much earlier. Webster's dictionary describes a journey as "an act or instance of traveling from one place to another." The road from motherhood to being a MoM (Mother of a Marine) was, for me, a long and painful one. March 23, 2015, marks the day I watched my little boy turn with a big-eyed grin and walk through a doorway to a world I would never experience. Only a mother would see the unspoken fear in his eyes, the quick questions of his decision, the squared shoulders of resolve, and the silent nod of an understood, "I love you, Mom." Those few steps took only mere seconds, yet when I replay them in my mind it's as if I see it all in slow motion. I freeze-frame each flash and soak it in. With every footstep that takes him further down the hallway, my heart collects each stomp. I wanted to stay still, stare at the door, and wait for him to return, but it would be 13 long weeks before I saw his face again. This is my journey.

I did not come from a military family, so it came as quite a shock when my teenage son expressed interest in the Marines. I assumed it was a phase, a chapter, a period that would pass quickly when something else caught his attention. When a recruiter called my house, I told him to never call again and hung up on him. I blocked his number from my cell phone. I raged at the audacity of military personnel visiting schools. I questioned how they could justify befriending these innocent children and persuading them to do something so tragic as to join the Armed Forces? I begged. I pleaded. I argued. I shamed. Finally, by graduation, he stopped talking about it. I successfully crushed the dream.

  Fast forward a few years to 2014. College did not work out well. He worked, sometimes at two jobs, but never seemed fulfilled. And I began to grasp the reality of what I may have done. He was bobbing frantically in a sea of uncertainty, and he would not talk to me about it. But why should he? I had shown him years before how sincerely I held his dreams. I have always been taught to believe in the power of prayer, and I began to do the only thing I knew to do: pray. I prayed that if the military was truly what God had planned for him, that I would begin to feel a peace about it.

  Just days later, my son told me one of his best friends had joined the Marines, and he was thinking about it as well. I asked him to pray with me, but to be open to the thought that if he didn't feel God was in the decision, he would be willing to change his mind, and in turn, I would do the same and be willing to change my mind if I felt God leading me differently. I had full confidence that if we were both sincere, God would not give us differing opinions. (I Corinthians 14:33 (NKJV), *"…for God is not a God of confusion…"*).

  That very night as I prayed and read my Bible, I came across a couple of verses I had underlined, and I didn't even remember them. In Galatians 4:15 (NIV), *"What has happened to all your joy?"* I couldn't help but cry as I remembered the cheerful little boy he once was; that gut-deep belly laugh that flowed from him. Galatians

4:19 (NIV), *"My dear children, for whom I am again in the pains of childbirth until Christ is formed in you."* I remember well the pains of childbirth; most mothers certainly do. I had prepared myself to labor in prayer just as in childbirth, to see Christ formed in his life. In the margin of my Bible, I had written the reference Psalms 18:34. I almost let it pass. But my eyes were drawn back to it, and I tried to think back to remember just when these passages had spoken to me before. My mind was blank. Somewhere one day long ago, God knew that I would need this confirmation, and at a time I don't even recall, He allowed me to mark my Bible in the very specific way I would need it for today.

I decided to turn to it just to see if it jogged any memories, and what I read stopped me completely.

> *"He trains my hands for battle." Psalms 18:34 (NIV)*

I began to sob. My hardened heart cracked, and I began to acknowledge the truths God was placing before me.

> *As for God, His way is perfect: the word of the Lord is flawless. He is a shield..." Psalms 18:30 (BSB)*

"I wasn't instantaneously on board. I feared what the road ahead would be like. But through the fear, I had a peace and a calm that comforted my terrified heart.

The Folder

He brought home the folder from the recruiter's office, and timidly handed it to me as he said, "It has some information you might be interested in." My mother's eyes teared up because for just a moment, I saw that dirty little chubby hand offering me a fistful of half-dead flowers through the back door, waiting patiently for my approval. I tried to accept his gift now with as much enthusiasm and gratitude as I did back then.

Although what he was placing in my hands was ripping my heart apart, I knew it was the most priceless thing he had to offer me: his hopes and his dreams of what he could one day become. I sincerely gushed over it, and his smile of relief was my assurance that I had responded appropriately. It was a full week later before I could bring myself to actually open the folder. As long as I left it closed, I could continue to pretend there was nothing to see. I would walk by the folder laying on my dresser, sometimes even stopping to stare at it, never touching it or opening it.

As the days passed, I knew he was right. There was information there I was not only interested in, but desperately needed. I quietly opened and explored the folder. I was stunned. This was real. This was happening.

There was a sticker for my car, but I would have to wait until he was actually a Marine to use it. There was information about recruit training, phone numbers for things I didn't understand, email addresses for people I didn't know, and pictures of smiling faces, handshakes, and salutes. My eyes froze on the photos of those successful Marines. I was in awe of the distinguished dress blues on those 'men.' How in the world could my little boy ever fill that uniform? Again, I cried.

Tears in a Bottle

He didn't leave for Parris Island for a few months, so we had some time to wait. I wish I could tell you that I was a wall of strength and a beam of joy. I tried to be. Oh, how I tried. But in the quiet of my room, I was a stream of continuous tears. I marveled at the verse in Psalms 56:8, (NLT) *"You have collected all my tears in your bottle,"* and wondered how many bottles He had collected for me. How huge must this room be in heaven to hold such a collection of a mother's tears? But I believe His Word, and once again, His Word is what sustained me. I began a journal to sort out my thoughts. What began as healing for me, quickly became a memoir to my son, with the hope of having it finished before he left.

I determined instead of making it a bitter sound-off for myself, to make it a cheerful diary of his childhood. I chose to not focus on my worries and my fears; I knew he would have enough of those on his own. So I filled it with funny tales of him as a little boy, my early-mommy dreams for him as a baby, things he once said or did that are fixed in my heart, everything I wanted him to know or hear but had never told him. In focusing on him and what he would need for recruit training, I found the strength and faith that I would need.

Day 2

I hope you can understand that my reluctance to accept this is not in ANY way because I think you are not capable. I KNOW you are! I have always believed in you. I've always known you were going to do great things. I just want them to be the things God has for you. And above anything else you do, making sure your eternity is sealed is the absolute most important. I don't want God to be someone you look to when all else has failed. I want Him to be your first thought, your Best Friend. I don't want you to make it to heaven by the skin of your teeth. I want it to be your ultimate goal—your ultimate hope. Above all, I want you to fall in love with Jesus. Everything else is secondary.

I love you always, Mom.

Day 3

Today has been a reflective day for me. I've been thinking about how things have been. You were the best baby. From the very beginning, I wanted you to be a boy. And I could not have made you more perfect if I created you myself. You were beautiful. A mop of black hair, round chubby cheeks, and dark eyes. You never cried. You were never bothered if your schedule was changed. If you fell asleep at home, in the car, in your bed, or at a store— it didn't matter. You just rolled with it. Whatever it was— you just rolled with it. As you got a little bigger, you loved your sister so much. You'd smile at anyone, but no one could make you belly laugh like your sister. And how she loved you! She'd line up her baby dolls, you along with them, and mommy all of you. And once again, you just rolled with it! Those years are some of the sweetest memories I treasure. I loved those years.

I typed, I babysat, but I did it all at home with you guys. You all were my everything, and if I ever had to go back in time, it would be to those days. They were my favorite. Know that I always have your back. Even when you're a Marine, miles away, know that your mom is covering you with prayer, protecting you the only way I can. When others look at you, they may see a strapping Marine, but I will always see a mop of black hair, those dark smiling eyes, and hear that deep belly laugh. I love you always.

Wishing I could stop time, Mom.

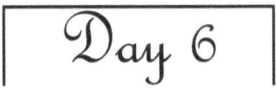

I believe in you. I believe in what you are doing and that you are following God's plan for you. But you can't do this without Him. Start every day by asking Him to be with you; by welcoming Him beside you wherever you may be. God will be with you, but He will never be where He isn't wanted or invited. Make Him a part of your daily routine, even if it's in the shower or when you lay down at night. Wherever it is, make Him welcome. My heart will always be with you, but all my love cannot protect you like the power of God in your life. I remind myself daily that He loves you even more than I do. When I have to leave you and let you get on that bus, I know you won't be going alone. He will be beside you the whole way.

    I love you always, Mom.

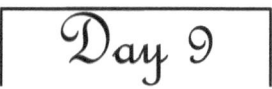

    Remembering the day you wanted to pray for the man in the car next to us because he couldn't afford a top for his car. (He was in a red convertible with the top down.) You've filled so many of my days with smiles.

    I love you always, Mom.

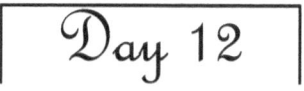

    The good thing about reading from all these websites is I know a little more what to expect. The bad thing about reading from all these websites is—I know a little more what to expect. As a mom, it has been my life's job to *make sure your "spirit" was never crushed. You don't know how many times you would be in trouble for something, and I'd leave you in your room praying I hadn't crushed your spirit with your punishment. And now to send you off to some island where their whole purpose is to crush your spirit is almost more than I can bear. I've spent an entire lifetime assuring you that you are a wonderful person with abilities and traits no one else comes close to. Now in 13 weeks, they are going to crush what I've spent a lifetime building. To know that you will feel like trash on the street is higher in value than you breaks my heart. And it makes me MAD! I know (supposedly) everything they do is for a reason, and everything is meant for your good in the end. But please, please, please, when you are at your lowest and feel like the gum on the bottom of someone's shoe, know that your mom thinks you are the greatest, the best! I'm so proud of you, and I'll be praying for you each day.*

    Remembering you are the BEST, Mom.

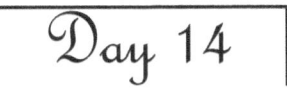

    I wrote a letter to you and your sister years ago. Part of it says this, "I wish I could keep you little forever; when all it takes to heal your hurts is a kiss and a band-aid. I dread the day when you realize that there are some

hurts "mommy's kiss" can't fix. I wish I could shield you from life's hurts, but I know that's not possible. Always remember how proud I am to be your mom. You are wonderful, and I love you with all my heart. No matter where you go, you're always as close as my heart."

You were only three years old when I wrote that, but I still mean every word. I may not be able to reach you or to get to you with a band-aid, but you're always in my heart. You are wonderful.

I still love you with all my heart, Mom.

Day 15

Another letter from November 1993:

"Ecclesiastes 9:10 (NIV), 'Whatever your hands find to do, do it with all your might." That kinda means to have passion about something. I hope I'm raising you to be passionate about things important to you. And I want you to know something: sometimes society doesn't portray being a mom as a very prestigious job or career. But to me, motherhood is my greatest honor. Don't ever think I've felt cheated out of a career. YOU are my career. You are my passion. It's my privilege to hear you call me "mommy." I wouldn't want it any other way. I hope one day you believe I was a mother "with all my might." I'm trying. Be strong. Be brave. Always love Jesus."

When I wrote this, I had no idea exactly what your hands would find to do. But I'm still praying you do it with all your might.

Be strong. Be brave. And love Jesus, Mom.

Day 16

From March 1994:

"Compassion is when you feel so sorry for someone that you have to do something to help them. Never forget to help others when you can. The other day we saw a man getting out of a car with a briefcase. You wanted to stop and help him because you thought he was living out of a suitcase. Never lose your compassion. Matthew 25:40 (NIV), 'The King will reply… whatever you did for one of the least of these… you did it for me.'"

Day 19

On those days when you wonder why your sister is having such a hard time with this, remember this. This is a letter I wrote to her when you were only 3 months old, and she was two.

This is your first Christmas with a new baby brother. I'm so proud of you! You're such a little mom. You love to help me take care of him. You love to sit on the floor with him and watch TV, and you want him to lay in the bean bag right beside you. Yesterday you sat on the floor eating Cheetos and watching The Little Mermaid. You, of course, had to stop eating when the songs played so you could dance around and sing. I had just fed your brother and changed him, then I laid him in the bean bag where he quickly fell asleep. You'd keep saying, "Wake up! It's the best part!" Over and over you'd want him to wake up. Finally, you gave up on keeping him awake and laid down in front of him to finish watching. You fell asleep holding onto his foot. I'm anxious to see how you grow up together. I wonder if you'll always want to "mother" him; make sure he doesn't miss the best parts of life. But I also wish I could keep you both little and with me, and never let you grow up. I know you will. But I hope you always find joy in simple things. Eating Cheetos, watching TV, and loving your little brother. May you both always love Jesus, because He always loves you. And so do I, Mommy.

Day 22

The time gets closer for you to leave, and I so wish I could slow it down. Each day I find new things to worry about. Surprised? I try to turn those worries into prayers. I pray your lungs will be strengthened, your headaches will be healed, your hand will be like new. I pray you find muscles you didn't even know you had, strength you didn't know existed, and a heart that never forgets how much you are loved. I've filled my head with positive thoughts, and my prayer is that at just the right moment, God will bring them to your memory.

Do you remember when you all were younger, and I used to give you scriptures to memorize? Your brother was little and didn't understand the importance of memorizing the Bible. I tried to explain it by telling him we may not always be able to have our Bibles with us, but if we had the verses in our hearts, God could remind us of the right scripture at just the right time. I told him when something bad happened, we'd remember the verse we needed in a snap! So he decided he wanted to pick his own verse to learn and didn't want to tell me what it was until he had it memorized. I agreed.

Not long after, we were on our way to church, going around a curve, and a tree had fallen across the road. We all screamed, I slammed on the brakes, and no one paid much attention in that moment to the words flying out of his mouth. It was only after everything calmed down, he said, "You were right, Mom. When something bad happened, that verse just popped in my head, and I said it!" I asked what his verse was, and he proudly quoted, "Noah tilled the land and planted a vineyard!" I'm not sure how that verse helped him in a bad situation, but it did. Who am I to question why or how?

My prayer is that somehow in these ramblings of a mother's heart, that the words, verses, memories, or thoughts that you need at any time will be sealed in the folds of your brain and released and reminded to you at just the right time. I'm praying that your mind is so crammed full of positive thoughts, happy memories, and sincere prayers, that there is not one speck of room for doubt or a thought of defeat to find room. You come from a long line of believers, and you can be sure you will have your own private army of prayer warriors on their knees for you daily. (Sorry for the Army reference, I couldn't figure out how to make it a Marine thing!) So soak in all these words of wisdom from your wise old mom!

I love you, so much more than you can realize, Mom.

Day 24

 I hope you remember this day I'm about to recall. It's probably one of my favorite memories with you. When you were in Kindergarten, you got out earlier than your sister. I'd pick you up, and we'd do things together. Usually, we would go to the park, get ice cream, or go to the store. This particular day we were going to the mall, but I had to get gas first. You were convinced you were a big boy and could help me get gas. So I let you. I let you hold the nozzle all by yourself. You were thrilled. Until you turned your head to tell me something and pulled the nozzle out of the car without letting go of the handle. You were covered in gas. I was covered. You screamed. I screamed.

 We came home, had to take showers and baths, and get rid of the clothes. That gasoline smell didn't leave us for days. When we finally got clean, we didn't have much time left but went to the mall anyway. Scrubbed and clean. We were walking in, and you never could resist a mud puddle. So when you got to it, you stomped your little clean foot in the middle of it. Neither of us realized how deep that puddle was, and mud flew ALL OVER ME! All over my clothes, my face, my hair. You looked as surprised as I felt. Your eyes were huge, just staring at me as dirty mud water dripped off my chin. For a few short seconds, it was as if time just froze. I can still see your scared little face, your clean red shorts, and red and white striped shirt. I yelled your name in my meanest voice, then stomped my foot into that puddle and splashed dirty mud water all over you too. I will never forget your laugh or the look on your face when you realized you weren't in trouble. We had more fun stomping in that little mud puddle than we would have ever had in the mall getting a cookie. We went back home filthy, but happy.

When I'm thinking about little things I'm thankful for, that day is definitely at the top of my list. So if one day, while you're on the island, you find yourself face first in the mud, remind yourself of happier mud days. And know that if someone is screaming in your ear, that is NOT the way Momma would have done it. But you are strong. You are brave. You've got this. I may not be there in person to sling mud back, but trust that I'm holding you in my heart. And trust that God's got your back. I loved to tell you when you were little: Be strong. Be brave. Always love Jesus.

I'm so proud of you, Mom.

Day 27

You have always had such a good heart and have wanted to help others. We used to keep little baggies in our car to give out to homeless people we'd pass on the street, and you loved it. We'd keep crackers, raisins, a bottle of water, and in the winter, we'd add a hand warmer and a pair of gloves. I'd type up Bible verses you and your sister would pick out, and we'd slip them in the bags too. You were always on the lookout for someone to give them to. One day we were coming to a red light where a homeless guy was standing, and I was hoping you wouldn't see him.

We were on our way to Krystal to splurge for the report cards you and your sister had brought home. I knew we didn't have any bags in the car for him, and I also knew I didn't have enough money to offer him food and feed us too. You insisted we feed him since we had no bag, and I reluctantly rolled down my window and told him if he wanted to walk to Krystal, we'd buy him some food. You were so excited! He walked over and met us there. I had decided to just not eat. You and your sister ordered kid's meals and ran to get a seat for us. When I asked him what he wanted, he told me to just get him whatever I was eating, then he asked the worker if it would be okay for him to use the restroom to wash his hands. She told him yes, he left, and I ordered only one meal since I didn't have enough money for two.

I knew she had overheard what he had told me he wanted, so she was rather confused as to why I was only ordering one meal. I saw in her eyes the split second she realized why, and I was so embarrassed. She went to the back for a second, and when she came back, she was smiling and said, "I was right. That meal you ordered is buy one, get one free tonight." I've never known Krystal to have a buy one, get one free meal, but just like always, God took care of us, and this time through a total stranger. It was all I could do to not cry right there in the middle of Krystal.

When he came to our table, he didn't sit down. He just stood looking, I'm guessing waiting for me to offer him the food so he could leave, but you pulled the chair out right beside you and told him to sit next to you. I assured him he was welcome, and he reluctantly sat down. You talked his ear off. You knew no shame, just came right out and asked him, "How come you're homeless?" We found out he was from Florida and had lost his job. He came to Georgia for work, lost that job, and couldn't find another one. He was too ashamed to go back home to his wife and kids with no job and no money, and that's how he ended up on the streets. We talked to him about Jesus and prayed with him before we left.

Even then, you liked the UT Vols. You were wearing a bright orange toboggan and wanted to give it to him. He told *you no, and you told him to take it because it was really warm on your ears, and his ears looked cold. He took it as tears began to slowly roll down his cheeks. He thanked you and put the hat on his head. We went outside, and you ran to the car, then came running back with your UT blanket Mamaw had made for you. You wanted him to have it too. He kept telling you he couldn't take your blanket. I told him to not deny a little boy the chance to be a blessing. He cried again as he hugged the blanket to his chest, and you were beaming. I was so proud of you. The man knelt down to you and stuck out his hand to shake your hand. But you ignored his outstretched hand, his filthy clothes, and just wrapped your little arms around his neck in a hug. He was stinky; he was dirty, but you never even noticed. You were the hands of Jesus that day. And most importantly, you were the heart of Jesus.*

You jumped in the car waving, and he told me what a great job I was doing raising such wonderful kids. To a single mom who questioned everything she ever did, those words were the sweetest ones I could hear. He said most people just stared at him. Few smiled. Mostly he was just ignored and not seen. He said the food was good, but what made the meal special for him was that somebody SAW him. And for one short evening, he almost felt as if he lived a normal life. I have no doubt that God showed up at Krystal that night, all because of your desire to help. For several days afterward, you'd ask to pray for him, and you would often ask, "I wonder what he's doing tonight?" I gave you the same answer each time and would tell you I didn't know. And each time you would smile and say, "I bet he sure is warm, though." You've always put other people first, so it shouldn't come as such a surprise to me that you want to be a Marine. After all, they only accept the best.

    Be strong. Be brave. Love Jesus. I love you, Mom.

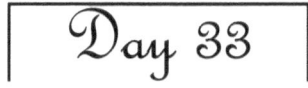

Day 33

    Well, I guess I need to find a way to wrap this up so you'll have time to actually read it before you leave. Just when I think I have told you everything there is to say, I can always think of something else. But it actually all boils down to this: I love you. I am proud of you. I will miss you. And you will constantly be in my prayers. I know this will be the hardest thing you have ever done. And it will be the hardest thing I have ever done too. I am trusting God to not only go with you and be with you but also to whisper in your ear the exact words you need at the exact moment you need them. I found this on the internet and it describes my feelings well.

My Heart Is On An Island (Author Unknown)

My heart is on an island
that seems oh so far away.
My mind is caught between two worlds,
yet it is here that I must stay.

This isn't just any island,
full of beaches and sunny scenes,

No, this is a place where courage lives,
and changes men and women into Marines.

Endurance, determination, and perseverance
are needed each and every day.
And all I can do is think of him,
send encouragement, and pray.

So I pray the Lord watches over him,
and helps him keep on going,
For he can't see my pride for him,
that others here are showing.

With every day that passes,
the feelings are even harder for me to hide.
I may be here, but my heart is not.
For it is at his side.

March 22, 2015

We dropped him off at MEPS (Military Entrance Processing Station) the night before he was to leave, and they put him in a motel room. We arrived back the next morning and met him. He was called in and out of offices, asked several times if he was joining of his own volition, and was he sure about his choice. There were others there with us from all branches. We were taken into a room and given some quick information and lunch, then back out to the waiting room again. Finally, they were all called into the room where they take the oath. We were allowed to go in also and given permission to take photos and videos.

> *I, _____, do solemnly swear (or affirm) that I will support and defend the Constitution of the United States against all enemies, foreign and domestic; that I will bear true faith and allegiance to the same; and that I will obey the orders of the President of the United States and the orders of the officers appointed over me, according to regulations and the Uniform Code of Military Justice. So help me God.*

It was a very sobering moment. There was no going back now.

Soon after, he walked quickly out of a room and said, "It's time." Things started happening so fast, I wasn't even sure what was going on. We were moved to a smaller waiting area for only Marine recruit families, and the recruits were lined up against the wall. With some quick words, they were put in a line with their backs to us to turn into the hallway and leave. With one look back, a slight smile, and a quick raise of his eyebrows, he was gone. His walk down the hallway took him to an underground parking garage where the bus was waiting for them. I wanted to stay there in case he came back. I wanted to run into the street to try to catch a glimpse of his bus leaving. The range of emotions was incredible. Eventually, everyone got up to leave the room in solemn silence. There were no cheerful smiles or friendly goodbyes to the other parents we had become acquainted with. There were no sighs of relief that it was over. Because all of us knew the unspoken truth: it had only just begun. We walked back to our car in silence. I was praying no one would speak to me because I was holding it together by a fragile thread. I knew one word would release an avalanche of tears that I would not be able to stop.

I honestly don't remember much of that day afterward. I suppose I was on autopilot and just getting through the day. I knew the next big step was waiting for the phone call. I had been told what it would be like. I had read other parents' experiences and thought I was prepared. I knew there would be yelling. I knew it would be loud and noisy. I knew exactly what he would be expected to say.

> *This is Recruit _____. I have arrived safely at Parris Island. Please do not send any food or bulky items. I will contact you in 7 to 10 days via postcard with my new mailing address. Thank you for your support. Goodbye for now.*

I had been warned that trying to yell back a quick "I love you" was more for my benefit than for his, as he probably would not be able to hear me anyway. I knew that some parents did not even receive the phone calls for different reasons. I prayed he would be able to remember my phone number. I desperately wanted that phone call. I had been told to expect it late at night, usually in the middle of the night, to keep them disoriented. I knew to answer any and all calls from unknown numbers just in case.

On March 23, 2015, at 8:03 PM, I got the unknown number call. I immediately answered, and was shocked to hear this hoarse whisper trying his best to yell as loudly and as quickly as he could. Just this morning, he had a full strong voice, but now this harsh raspy tone shouted as best as possible, "This is Recruit Davis. I have arrived safely at Parris Island."

First of all, my son had never yelled at me before. And the voice; I had never heard his voice more strained. I was in shock. I remembered at the last second to scream, "I LOVE YOU," and then the line was silent.

I sat there unmoving for the longest time, still holding the phone to my ear. Then little by little, continuous lines of tears started rolling down my cheeks. There were no sounds. If you didn't see the wet pool forming in my lap, you might not have even known I was crying. It was as if someone had turned on a spigot to my eyes, and I could not turn it off. I don't know how long I sat there. As a single mom who raised my kids on my own for their toughest years, I had been through a lot. But nothing had ever broken my heart like that phone call. The next few days were a blur. It was a strange feeling, really, knowing that he was not at home, but there was no wondering about where he was or what he was doing. One thing was certain, he was on an island, and he wasn't getting off until he was a Marine.

The Routine

The next hurdle was waiting for the letter giving me his address. I began to go ahead and write letters to him, even putting them in the envelopes and numbering them in the order I wrote them. All I needed was to get the address, and I could drop them in the mail.

Finally, the first letter came. It was just a form letter, nothing personal to it. Still, it gave me the address, and I immediately started mailing letters to him. My daughter and I got into a pattern of writing him every other day on opposite days, so one of us was continually putting a letter in the mail to him each day. His letters came less often, and sometimes they were on little scraps of torn paper that looked like trash he had saved.

We started sending him stamped envelopes with our addresses already printed on them. That way, all he had to do was put a quick note in and seal it. He had so little time. I even started typing out questionnaires for him and gave him multiple choice answers he could choose. Or questions he could quickly answer on that sheet and send back to me. We printed off pictures of his dog, a picture of our house with a Marine wreath on the door, pictures of positive affirmations, scriptures, song lyrics, anything we could think of to keep his spirits up.

He had told me before he left, to count his time gone in Sundays instead of days or weeks. He would be gone 13 Sundays. So I made a paper chain with 13 links, and each Sunday I removed one link. I would send him pictures of the shrinking chain.

We poured over the websites to try to find glimpses of him. It is true, they all start to look alike very quickly. I started looking for little things to try to differentiate him from someone else. Was that his ear? Does his finger have that crook to it? And finally, there was a picture, and there was no mistaking; it was him. He looked so thin, and so young, yet so mature. What was happening to my little boy? That rifle was half the size of his body, yet he carried it with ease.

So often God placed verses in front of me for the exact place I was. I knew the Bible was going to be my source of strength. One of the first verses I found was in Psalms 69:3 (CSB), "I am weary with my crying out; my throat is parched." It reminded me of the phone call and how, after just a few hours, his voice was gone, his throat was obviously parched, and I could only imagine how weary he must have been.

Second to my Bible, the Matrix became my best friend. The Matrix is a grid showing what they were expected to do each day. I would try to stay a week ahead, watching what was coming up for him, and write to him about it the week before. I don't know how closely they stayed to the Matrix, but it was a good grounding tool for me and helped me to feel as if I were in the loop somehow. Every day I poured over the Bible, and once again, God pointed me to the exact scriptures I needed. Always the verses were spot on, but sometimes they were so comical, I had to laugh. I am confident that God has a sense of humor, and He knew the exact moments I needed to loosen up and smile.

One of the first things he did in the early days was to meet his Drill Instructor, and I came across this verse:

> Psalms 18:48 (NASB), "Rescue me from the violent man."

I marveled at David's words and wondered if maybe David had a DI himself.

> Psalms 57:4 (NIV) "I am in the midst of lions; I am forced to dwell among ravenous beasts—men whose teeth are spears and arrows, whose tongues are sharp swords."

I prayed my son would have the heart of David in the face of those men with sharp swords for tongues. I marked these verses from Psalms and prayed he would remember them when he met his DI.

> Psalms 56:4 (NIV), "In God I trust. I will not be afraid. What can mere mortals do to me?"

> Psalms 59:7-8 (ESV), "There they are, bellowing with their mouths...but you, O Lord, laugh at them."

During Bayonet Techniques, this verse brought comfort:

> Psalms 22:20 (NIV), "Deliver me from the sword..."

I had been warned of the "hurricane" that came during recruit training. We were told to send gallon size baggies so he could begin "preparing" for the storm.

Once again, the Word brought a smile. I sent this verse along with the baggies:

> Psalms 57:1 (ESV), "Be merciful to me, O God, be merciful to me, for in you my soul takes refuge; in the shadow of your wings I will take refuge, 'til the storms of destruction pass by."

The Confidence Course is a course of obstacles they are required to pass. That week I sent him this verse,

> Psalms 18:29 (ESV), "By my God I can leap over a wall."

Before he left, I was confident Swim Week would be easy for him. He had been a river guide and was an excellent swimmer, but like everything else, when the time came, I worried even about swim week. All the what if's and horror stories others told got in my head. But again, these words brought peace.

> Isaiah 43:2 (NIV), "When you pass through the waters, I will be with you."
>
> Psalms 69:1 (NIV), "Save me, O God! For the waters have come up to my neck."
>
> Psalms 18:16 (ESV), "He drew me out of many waters."

The gas chamber also brought on terror. He had struggled with asthma earlier, and I was worried about the effects the gas chamber would have on his lungs. I identified with Daniel:

> *Daniel 10:17 (NIV), "My strength is gone, and I can hardly breathe." But then came the rest of the story in verses 18 and 19 (CSB), "The one who looked like a man touched me and gave me strength. And he said, 'You who are treasured by God, don't be afraid. Peace to you; be very strong.'"*

My sister told me, and I shared with my son, "The same God who kept Daniel in the lion's den will be with you in the gas chamber. You are treasured and strong!" (Consequently, when he got home he said the gas chamber was one of the easiest experiences.)

For Team Week:

> *Galatians 6:2 (ESV), "Carry one another's burdens."*

On Sundays, it was a comfort to know that when I was going to the house of God, he had the opportunity to do the same. A short escape from the routine, a brief distraction from the DI, a break from the yelling and constant demands. Miles apart, yet worshiping the same God.

> *Psalms 95:6-7 (NLT), "Come, let us worship and bow down; let us kneel before the Lord, our Maker."*

So often, missing him seemed like physical pain. It wasn't just the missing his presence. It was the knowledge that he was going through things that I could not help him through. It wasn't something I knew how to explain to others. Some well-meaning people would try to relate by comparing it to when their child left for college or moved away from home. I would just smile and nod because if you've never had a child leave for recruit training, you just are not going to understand. There were some days I would just go lay on his bed, or go to his closet and pull out a shirt, all in the hopes of it seeming as if he were there at home. And then I came across this verse from Genesis 27:27 (NIV), *"Ah, the smell of my son is like the smell of a field that the Lord has blessed."*

When I think of a smell of a field, I think of wildflowers, and butterflies, and sunny days. I can remember many times (especially during those middle school years) I would not have considered the smell of my son like the smell of a field! But as I reflected on that verse, I realized the smell of our kids, our babies, never leaves a mother's memories. Just for a moment, if you close your eyes and breathe in, you can remember. That tiny bundle with long dark lashes laying on puffy pink cheeks that you carried home from the hospital, that little lotioned fist wrapped around your finger, that chubby toddler you snuggled in a towel fresh out of the dryer, that little kid straight from the playground you picked up from school, that not so gentle scent of a growing pre-teen, that youth with too much cologne on a first date, and that empty, abandoned bed, the stale closet smell of the aspiring young Marine you are missing; the smell of little boys never leaves a mother's memories. Ah, the smell of my son.

When my kids were smaller, I would try to give them easy ways to remember the Word. I wanted it to be rooted in their young minds, words they could always remember. I would tell my son often, "Be strong, be brave, and always love Jesus." I borrowed the words from Psalms 31:23-24 (CSB),

> "Love the Lord, all his faithful ones…Be strong, and let your heart be courageous…".

I began to use these words again in letters to him during recruit training. Be strong. Be brave. And love

Jesus. I have always loved it when God's Word confirms Itself in ways you don't expect, and so often that happened as I began to study and look for words of comfort. Joshua 1:9 (CSB),

> *"Haven't I commanded you, be strong and courageous? Do not be afraid or discouraged, for the Lord your God is with you wherever you go."*

My continuous prayer throughout recruit training was inspired by Psalms 91:10-11. I constantly prayed for no harm to come to him, no sickness to come near his barracks, and that angels would protect him in all his ways. I often reminded him in letters, and myself in my thoughts, that recruit training was a long night, but joy comes in the morning, based on Psalms 30:5. I prayed over him,

> *"Lord, show him your favor, and make him stand like a strong mountain."*

There were so many verses I prayed over him daily, sometimes hourly. "Lord, please open his eyes and let him see." I had included scriptures in every letter I sent him, and I prayed whatever he needed would be brought to his memory at just the moment he needed it.

> *Isaiah 41:10 (CSB), "Do not fear, for I am with you; do not be afraid, for I am your God. I will strengthen you; I will help you; I will hold on to you with my righteous right hand."*

> *Isaiah 46:4 (CSB), "I have made you, and I will carry you; I will bear and rescue you."*

Psalms 31:24 (CSB), "Be strong, and let your heart be courageous."

Luke 21:36 (CSB), "But be alert at all times, praying that you may have strength to escape all these things that are going to take place."

Exodus 14:13-14 (CSB), "Don't be afraid. Stand firm and see the Lord's salvation that he will accomplish for you today... The Lord will fight for you, just be still."

Ps 69:13-15, 19-20, 29-30 (CSB), "My prayer is for a time of favor... Rescue me from the miry mud; don't let me sink...Don't let the Pit close its mouth over me. You know the insults I endure—my shame and disgrace. Insults have broken my heart, and I am in despair. I waited for sympathy, but there was none; for comforters, but found no one."

Psalms 55:17 (CSB), "I complain and groan morning, noon, and night, and he hears my voice." (Believing God hears your prayers even if you're hoarse)

Psalms 16:6 (CSB), "The boundary lines have fallen for me in pleasant places; indeed, I have a beautiful inheritance." (Inheritance is your Christian heritage. You have a long line of believers praying for you.)

Psalms 3:5-6 (CSB), "I lie down and sleep; I wake again because the Lord sustains me. I will not be afraid." (Praying you sleep well tonight.)

The Crucible

The Crucible is the final test for the soon to be Marine. It is 54 hours, 45 miles of marching, carrying at least 45 pounds of gear. It also includes food and sleep deprivation. We had done the crucible candles, had many people praying, and knew about when their Crucible started. It was a restless couple of days for me, and I knew it was nothing compared to what they were enduring. We had a prayer chain and had someone praying every minute of their 54-hour Crucible.

My time slot was Saturday 0200-0300, but my actual prayers were every minute I was awake. There were many verses God pointed me to during those two days. Psalms 91:11-12 (ESV), *"He will command His angels…to guard you in all your ways…lest you strike your foot against a stone."* But I claimed Psalms 91:14-16 (ESV) as what I called a 'Crucible Promise,' *"Because he holds fast to me in love, I will deliver him. I will protect him because he knows my name. When he calls to me, I will answer him. I will be with him in trouble. I will rescue him and honor him."* That word leaped off the page to me: honor. One of the three Marine Corps Values: honor, courage, commitment. I began to keep a prayer log of his 54-hour journey.

2 hours in

Isaiah 46:4 (CSB), "I have made you, and I will carry you; I will bear and rescue you."

4 hours in

Isaiah 41:10 (CSB), "Do not fear, for I am with you; do not be afraid, for I am your God. I will strengthen you; I will help you; I will hold on to you with my righteous right hand."

16 hours in

Psalms 118:13-14 (CSB), "They pushed me hard to make me fall, but the Lord helped me. The Lord is my strength and my song; he has become my salvation."

23 hours in

Psalms 118:5-6 (CSB), "I called to the Lord in distress; the Lord answered me and put me in a spacious place. The Lord is for me; I will not be afraid."

24 hours in

Luke 21:34, 36, "Be on your guard, so that your minds are not dulled...But be alert at all times, praying that you may have strength to escape all these things that are going to take place..."

25 hours in

Luke 22:32(CSB), "But I have prayed for you that your faith may not fail. And you, when you have turned back, strengthen your brothers."

29 hours in

Galatians 6:9 (CSB), "Let us not get tired of doing good, for we will reap at the proper time if we don't give up."

34 hours in

Psalms 71:20-21 (CSB), "You caused me to experience many troubles and misfortunes, but you will revive me again. You will bring me up again... You will increase my honor and comfort me once again."

36 hours in

Psalms 73:26 (CSB), "My flesh and my heart may fail, but God is the strength of my heart, my portion forever."

39 hours in

I sent song lyrics to him before the Crucible and was praying that God would remind him of the lyrics and just the right time.

Psalms 77:6 (CSB), "At night I remember my music."

54 hours in

The Crucible should be finished. He should be arriving at the Iwo Jima Memorial, where he will be presented with his eagle, globe, and anchor, and called a "Marine" for the first time in his life. I am crying tears of joy, relief, and hope. God graciously laid this promise right in plain sight for me to see, Psalms 89:19-24 (CSB), "I have granted help to a warrior...I have anointed him...My hand will always be with him. My arm will strengthen him. The enemy will not oppress him...I will crush his foes before him... My faithfulness and love will be with him..."

If all went well, I would get a phone call tomorrow. It would be the first time I had heard his voice since the fateful call when he got to the island twelve and a half weeks ago.

I didn't even go to church that Sunday, for fear I would somehow miss the call. And then, finally, "Hey Ma, I made it!" What a precious sound and an unforgettable moment. That tired, weary, hoarse voice had transformed into this strong voice I had not heard in so long. I don't even remember what all we talked about. I remember wondering what all he left unsaid. I knew there were things he went through on that island that he would never share with me. When we hung up, I felt like I had just become a mother all over again to a whole platoon of young men that I may never meet. When I sat down with my Bible that evening, once again, the words did not disappoint.

> *Genesis 48:15-16 (CSB),* "*The God before whom my fathers walked, the God who has been my shepherd all my life to this day…may he bless these boys…*"

PLT (Platoon) 3040 felt like my family.

The Reunion

      The day finally came for us to begin our trip to the island. I was overjoyed. We had everything lined up; rental car, house reservations, insurance cards in the glove box, identification for everyone. We made it to Parris Island and still had a day to wait before we could see him. As we drove over the long road to take us to the Island, there was water on each side. I tried to imagine what it had felt like for him to be riding on the bus on this long stretch. Was he scared? Was he excited? Was he heartsick? Was he regretting his decision? We saw the yellow footprints he would have stood on when he first got there, and I tried to envision which ones he would have been at. I wondered if he was as terrified then as I still felt for him now.
      We drove around to see this place that had been his "home" for the last 13 weeks. We wandered around, picked up pictures that had been made, and took in all the sights. We tried to look for a glimpse of him at every opportunity. I had heard the stories of how they all would look alike, and it would be difficult to pick your own child out from the crowd, and I had dismissed that thought. I was confident that I could pick my child out anywhere from anybody. And I was wrong. They looked

alike. They dressed alike. They walked alike. They moved as one. Feet in perfect rhythm with the others. At attention, their fingers touched the exact same point on their foreheads. At ease, their hands folded identically behind their backs. Their toes pointed out at the same angles. Their covers never left their heads, but if they had, we would have seen identical haircuts. To say, "spotting him was difficult," was an understatement. At one point during the day, we were sitting beside a family who had spotted their Marine. The mother was crying, exclaiming how handsome he looked, how much he had changed, how proud she was; then in the middle of her crying, she abruptly paused, and announced, "Oh wait, that's not him!"

    I took pictures of everything until we came upon "the pit." I could only speculate what he endured there. We had already had some exposure to the fire ants, and I assumed they were even worse in that sand. I tried to see everything through his eyes but quickly realized whatever I imagined, it had probably been worse.

    The next morning was the Moto Run. Bright and early before the sun was up, we gathered as a family in front of the Iwo Jima Memorial. We had made signs and posters in hopes of catching his attention and showing our support. Not long after the sun came up, we heard the cadence getting louder and louder as they got nearer to us. And suddenly, there he was in front of us. Head turned just quick enough for me to snap a picture, with just the hint of a smile that promised, "I'm almost finished." We got to yell and scream at him for the first time. We had laid eyes on him but still no contact with him. After the Moto Run, there was a Liberty Ceremony and Family Day. As parents, we were given the option to attend a quick class, where we were given some instructions on what things would be like. We were reminded that our new Marines had not been touched for 13 weeks. No hugs, no warm handshakes, no pats on the back. They had no cell phones, no newspapers, no televisions, no contact with the outside world. It would be a shock for their system to be back in real life. For 13 weeks, they had been told what to do, where to do it, and how quickly to get it done. They were told when and if they could shower or go to the bathroom. They were told when to eat and how rapidly to get it down. We were cautioned that our jubilant reunion could be stunning for them and that it was best to let them set the pace for it. We were instructed to give them directions for plans instead of giving them options when we got back home; to tell them where we were going and at what time, instead of asking what they wanted to do; to let them gradually fall back into the routine of making their own plans, or it could be overwhelming.

      I don't remember anything that was said during the ceremony. I just remember rows and rows of uniforms, standing straight as an arrow, with hands clasped behind their back. I marveled that no one moved, coughed, or fell over from sheer exhaustion. At the end of the ceremony, they were released for liberty. The floor flooded with families racing to their loved ones. I could not get to him fast enough. Then suddenly, there he was in front of me. The little baby I had rocked to sleep, the toddler I had chased after, the little boy I had laughed with, the teenager I had fought with, the starry-eyed youth with the slight smile and raised brow, was suddenly in front of me, embracing me. I had not seen him in three months, and what a transformation. He was skinnier. He looked so exhausted and worn out. Tears mingled with laughter as we all took turns getting our hands on him as if to assure ourselves he was actually there. We were able to spend a few hours together. We had a picnic overlooking the water. It was conflicting to look out over that peaceful water and remind yourself of what the last three months had held for him. Staring across that water, I considered how many parents had stood just where I was and wondered what was in store for their Marine. And more importantly, how many Marines had stood there wondering what was in store for themselves.

During the picnic, I noticed him occasionally walking with a limp. We found out the skin on his feet was completely shredded. He had so much damaged and blistered skin, every step was excruciating. I could not fathom how he had managed to stand for so long during the ceremony, or run in the Moto Run, or even walk at all. He had been given the option of not walking for graduation the next day because of the condition of his feet, but after all he had been through to get there, he was not about to miss the honor of marching with the graduating class. He showed us around the base and where his quarters were. I marveled at how he carried himself, his posture, his stance. I wondered if the difference was from how long it had been since I had seen or talked to him, or from the experience itself. All too soon, we had to let him go. We wouldn't see him again until the graduation ceremony the next day.

There was much waiting. We were soon to learn that in the military, your time is not really your time at all. Every event made clear the motto, "Hurry up and wait." Knowing we would not be able to talk to him, we still chose to wait there while they stood in formation to leave. We noticed the ones standing directly in front of us were falling asleep while still standing. It became comical and pitiful at the same time to see them so exhausted.

Graduation took place outside, June 19. June at Parris Island is unbearably hot. I stood there on the Parade Deck, waiting and watching. I was speechless at the myriad of emotions. There were people there from all walks of life, differing economic classes, all races, all sizes, all colors. Yet we were there unified as one family, with one purpose: to celebrate. Looking around, it was obvious that just getting to the island was a sacrifice for some families. I wondered what they had given up the last three months to be able to afford the trip to be there.

It occurred to me that there were most likely some Marines in that graduation ceremony that had no family there at all. I could not imagine not being able to be there to experience this, and I silently thanked God for the undeniable gift of this day. Graduation finally began, and it was a deeply emotional experience. It was more than just being proud of him, although I was extremely proud. I was proud to be an American. I have always been grateful to be an American, but this exceeded just being grateful. Things are seen and felt differently when there is a personal stake in the responsibility of defending this great nation. "Boots on the ground" means something entirely different when you realize you have flesh and blood involved.

It was uncomfortable to realize I had never thought of this before. The sacrifice did not seem so great until my son was the one willing to make it. What a cost to all the many parents and families before us, and yet I had never given them consideration. When you come face to face with your own selfishness and realize all you have taken for granted, and then look into the young faces of these brave men and women who are willing to literally stand up for your rights, it is a sobering and humbling experience. To watch the reverence radiate from their faces as the American flag was raised; to see the immediate snap to attention, the squared shoulders, the perfected salute when The National Anthem rang out; to witness the respect displayed when officers walked past; I soaked the experience in like a sponge. And I walked away with a little more reverence, a little more attention, a little more respect, a little more humility, and a whole lot of admiration and pride.

After graduation there was, of course, much picture taking and celebration. Perhaps one of the most moving moments for me was when my Marine presented me with a Marine Mother's coin. He called me over before I realized he had someone videoing the whole exchange. He put his arm around my neck and held his hand out to me. As he released the coin into my palm, my mother's eye quickly saw that little boy with dirty fingers giving me a frog. The body may have grown up, but the grin was unmistakable. The squinting eyes as he smiled, this was my boy. The back of the coin says, "Mothers, it is because you protected them that they are willing and able to protect you." And for one short moment, all the second-guessing myself, all the wondering if I had done anything right, all the stigma that comes from being a single mother just vanished. And I savored it.

Then just like that, we had come full circle. That little tiny baby with a mop of black hair, round chubby cheeks, dark eyes, and a deep belly laugh; that little one that I would lay down my life for, had somehow grown up into the strapping Marine that I had seen in the folder he gave to me months before, one that was willing to lay down his life for us.

Had it really been less than a year ago? It seemed like a lifetime ago. I had no idea where the journey would take us from there, but I had no doubt he was in God's hands, and so was I. We had both grown up during recruit training in ways I had never imagined. Different events, different tests, but a journey that brought us to the exact same outcome: Pride.

He had done it. And so had I.

Semper Fidelis.

www.ingramcontent.com/pod-product-compliance
Lightning Source LLC
Chambersburg PA
CBHW030202100526
44592CB00009B/399